I0466582

HOW TO DO PLUMBING

A COMPREHENSIVE GUIDE TO STARTING, LICENSING, MARKETING, AND GROWING A SUCCESSFUL PLUMBING SERVICE COMPANY WITH EXPERT TIPS AND STRATEGIES

Dukuwic ghawane

TABLE OF CONTENT

INTRODUCTION TO PLUMBING

Plumbing is an vital infrastructure that ensures the secure and efficient distribution of easy water for consuming, cooking, sanitation, and the removal of wastewater. It contains a network of pipes, fixtures, and fittings designed to supply water into buildings and remove waste to remedy centers.

From historic aqueducts to modern urban plumbing systems, the sphere has advanced extensively, integrating superior technologies to decorate efficiency and sustainability. Understanding fundamental plumbing concepts, renovation practices, and protection measures is important for making sure reliable water supply and protecting public fitness.

IMPORTANCE OF PLUMBING IN ORDINARY LIFE

Plumbing performs a important position in regular existence,

1. Access to Clean Water: Plumbing systems ensure get right of entry to to secure and clean water for ingesting, cooking, bathing, and hygiene. Without dependable plumbing, acquiring potable water would be tough and potentially unsafe.

2. Sanitation and Hygiene: Proper plumbing allows the efficient removal of wastewater and sewage from homes and buildings, preventing the spread of illnesses and retaining hygienic situations.

3. Comfort and Convenience: Plumbing structures offer convenience by handing over water at once to faucets, showers, toilets, and home equipment like washing machines and dishwashers. This enhances

daily residing requirements and performance in family responsibilities.

4. Environmental Sustainability: Modern plumbing technologies contain water saving fixtures and sustainable practices that make a contribution to holding water assets and reducing environmental effect.

5. Public Health Protection: Well maintained plumbing systems save you contamination of ingesting water and make sure that wastewater is safely disposed of, safeguarding public fitness by using lowering the threat of waterborne sicknesses.

6. Economic Benefits: Efficient plumbing structures contribute to assets price and marketability, as homes and buildings with dependable plumbing infrastructure are more applicable and practical.

BASIC PRINCIPLES AND ADDITIVES OF PLUMBING SYSTEMS

Basic concepts and components of plumbing structures shape the muse for knowledge how water is added and wastewater is eliminated in homes.

BASIC PRINCIPLES OF PLUMBING SYSTEMS

1. Water Pressure and Flow:

- Pressure Regulation: Ensures constant water strain all through the device, typically regulated by using pressure reducing valves (PRVs).

- Flow Dynamics: Understanding how water flows through pipes and fittings, affected by elements like pipe diameter and format.

2. Gravity and Ventilation:

- Gravity Systems: Utilizes gravity to move wastewater downhill thru drainage pipes to sewer strains or septic tanks.

- Ventilation: Allows air into the plumbing device to save you vacuum formation, aiding within the green waft of wastewater and preventing traps from siphoning dry.

3. Cross Connection Control:

- Prevents contamination by way of making sure that potable (ingesting) water structures are separated from non potable systems (e.G., wastewater).

COMPONENTS OF PLUMBING SYSTEMS

1. Pipes:

- Materials: Common materials encompass copper, PVC (polyvinyl chloride), PEX (gorelated polyethylene), and galvanized metallic.

- Sizes and Types: Vary relying on the software (e.G., deliver lines, drainpipes) and local constructing codes.

2. Fittings and Connectors:

- Types: Elbows, tees, couplings, and adapters facilitate the connection and path adjustments of pipes.
- Materials: Matched to the pipe material for compatibility and sturdiness.

3. Valves:

- Types: Gate valves, ball valves, and take a look at valves manipulate the waft of water, isolate sections of the device, and prevent back flow.
- Location: Found at strategic factors along with essential deliver traces, furniture, and home equipment.

4. Fixtures:

- Examples: Sinks, lavatories, showers, and bathtubs are related to the plumbing device to receive water and put off wastewater.
- Efficiency: Modern furnishings consist of water saving functions to

preserve water with out sacrificing performance.

5. Traps and Drains:

- Traps: P fashioned or S fashioned fittings beneath sinks and drains that maintain water to prevent sewer gases from coming into the constructing.

- Drains: Pipes that bring wastewater away from furniture to the main drainage machine or septic tank.

6. Water Heaters and Appliances:

- Water Heaters: Supply hot water for bathing, cleaning, and cooking.

- Appliances: Washing machines, dishwashers, and fridges with water dispensers require plumbing connections.

CHAPTER 1: TOOLS AND MATERIALS

ESSENTIAL EQUIPMENT FOR PLUMBING OBLIGATIONS (E.G., WRENCHES, PIPE CUTTERS)

COMMON MATERIALS (E.G., PIPES, FITTINGS, SEALS)

Common materials utilized in plumbing for pipes, fittings, and seals encompass quite a few alternatives applicable to different desires and packages:

PIPES:

1. Copper:

● Advantages: Durable, proof against corrosion, suitable for both hot and bloodless water, and has a protracted lifespan.

- Common Uses: Water deliver strains, especially in residential and commercial buildings.

2. PEX (Cross connected Polyethylene):

- Advantages: Flexible, easy to put in (regularly using compression or crimp fittings), immune to scale buildup and freeze harm.
- Common Uses: Water deliver traces, specifically in retrofit and new construction.

3. PVC (Polyvinyl Chloride):

- Advantages: Lightweight, inexpensive, immune to corrosion and chemicals.
- Common Uses: Drainage, vent structures, and every now and then water supply strains (generally cold water).

4. CPVC (Chlorinated Polyvinyl Chloride):

- Advantages: Similar to PVC however can face up to higher temperatures.

- Common Uses: Hot water supply strains, especially in residential packages.

5. Galvanized Steel:

Advantages: Strong, long lasting.

Disadvantages: Prone to corrosion over time, now much less generally utilized in prefer of extra corrosion resistant substances.

- Common Uses: Water supply traces, historically used in older homes.

6. PE (Polyethylene):

- Advantages: Flexible, proof against chemical substances.

- Common Uses: Irrigation systems, underground water deliver strains.

FITTINGS:

1. Brass:

Advantages: Durable, corrosion resistant, appropriate for high stress packages.

Common Uses: Various plumbing programs, such as fittings and valves.

2. Copper:

   Advantages: Matches properly with copper pipes, durable.

   Common Uses: Compression fittings for copper pipes.

3. CPVC and PVC:

   Advantages: Lightweight, easy to put in.

   Common Uses: Solvent welded fittings for PVC and CPVC pipes.

4. PEX:

- Advantages: Crimp or push fit fittings for smooth set up without soldering or gluing.

- Common Uses: PEX specific fittings for PEX pipes.

 SEALS AND GASKETS:

1. Rubber (Nitrile or EPDM):

- Advantages: Resilient, right sealing residences.

- Common Uses: Sealing pipe joints, valves, and fixtures.

2. Silicone:

●   Advantages: Heat resistant, flexible.
Common Uses: High temperature programs, together with around water warmers.

3. Teflon (PTFE):

●   Advantages: Chemically inert, used for sealing threaded pipe connections.

●   Common Uses: Thread seal tape (Teflon tape) for pipe threads.

CHAPTER 2. SAFETY PRECAUTIONS

IMPORTANCE OF SAFETY IN PLUMBING PAINTINGS

PROTECTIVE EQUIPMENT AND SECURE WORKING PRACTICES
Protective equipment and safe running practices are important in plumbing to save you accidents, make sure nonpublic protection, and adhere to occupational fitness requirements.

PROTECTIVE GEAR

1. Eye Protection:

● Safety Glasses or Goggles: Protect towards particles, splashes, and chemical substances when reducing pipes, the use of electricity equipment, or handling chemical compounds.

2. Hand Protection:

- Work Gloves: Prevent cuts, abrasions, and burns while coping with sharp or hot materials, tools, and chemical compounds.

3. Foot Protection:

- Steel Toe Boots: Protect against falling objects, heavy substances, and sharp gadgets on production websites or in doubtlessly dangerous environments.

4. Respiratory Protection:

- Respirators or Masks: Necessary whilst operating in dusty environments, with chemicals, or in confined spaces where air high quality may be compromised.

5. Head Protection:

- Hard Hats: Required on creation sites to defend in opposition to falling objects, bumps, and head injuries.

SAFE WORKING PRACTICES

1. Tool Safety:

- Proper Use: Use tools efficaciously and for their supposed purpose, following manufacturer instructions and protection guidelines.

- Maintenance: Keep tools in proper condition, frequently analyzing for wear and damage.

2. Ladder Safety:

- Stable Placement: Ensure ladders are on stable ground and secure earlier than use.

- Three Point Contact: Maintain 3 factors of contact (hands and one foot or two feet and one hand) while hiking.

3. Chemical Handling:

- Read Labels: Follow commands and warnings on chemical containers.

- Ventilation: Ensure a ventilation when running with chemical compounds to prevent inhalation of fumes.

4. Electrical Safety:

- Shut Off Power: Turn off electrical energy before working on plumbing furnishings or near electric wiring.

- Use GFCIs: Use Ground Fault Circuit Interrupters to prevent electric shocks in moist environments.

5. Confined Space Safety:

- Permit Required: Obtain vital lets in and observe techniques for coming into limited spaces.

- Ventilation and Monitoring: Ensure proper ventilation and nonstop monitoring of air best.

6. Personal Hygiene:

- Hand Washing: Wash palms thoroughly after managing chemicals, sewage, or contaminated substances.

- Protective Clothing: Change out of labor clothes directly to keep away from sporting contaminants domestic.

7. Team Communication:

- Safety Briefings: Conduct pre job protection briefings to speak about dangers, safety protocols, and emergency methods.

- Emergency Response: Know the vicinity of emergency exits, first useful resource kits, and emergency contacts.

CHAPTER 3: UNDERSTANDING PLUMBING SYSTEMS

TYPES OF PLUMBING SYSTEMS (E.G., WATER SUPPLY, DRAINAGE)

LAYOUT AND LAYOUT ISSUES

Layout and layout concerns in plumbing are vital for making sure efficient functionality, toughness, and compliance with constructing codes.

GENERAL LAYOUT CONSIDERATIONS

1. System Functionality:

- Separation of Systems: Ensure clear separation between potable water deliver, wastewater drainage, and vent structures to prevent contamination.

- Efficiency: Design layouts that limit the length of pipe runs and reduce the variety of fittings to optimize water waft and pressure.

2. Accessibility:

- Serviceability: Ensure that plumbing furnishings, valves, and clean outs are easily available for upkeep and repairs.

- Clearance: Provide good enough clearance around fixtures, pipes, and device for ease of set up and destiny renovation.

3. Space Utilization:

- Fixture Placement: Strategically location fixtures such as sinks, bathrooms, and showers to optimize area and functionality within rooms.

- Utility Areas: Designate areas for water warmers, water softeners, and different system, considering air flow and drainage necessities.

DESIGN CONSIDERATIONS

1. Pipe Sizing and Materials:

- Flow Rates: Calculate required pipe sizes based on predicted waft fees and stress drops to make certain adequate water deliver at some stage in the gadget.

- Material Selection: Choose suitable pipe materials (e.G., copper, PEX, PVC) based totally on elements like water great, temperature, and compatibility with different machine components.

2. Ventilation and Drainage:

- Venting Requirements: Provide sufficient venting to prevent traps from siphoning and ensure proper drainage of wastewater.

- Slope: Maintain right slope for drainage pipes to facilitate the green elimination of wastewater and save you standing water and blockages.

3. Pressure Regulation:

- Pressure Control: Install pressure decreasing valves (PRVs) wherein necessary to adjust water stress and defend plumbing furnishings from damage.

- Back flow Prevention: Incorporate back flow prevention gadgets to shield against infection of potable water substances.

COMPLIANCE AND SAFETY

1. Building Codes and Regulations:

- Local Codes: Adhere to neighborhood building codes, plumbing standards, and policies governing pipe materials, fixture installation, and safety practices.

- Permits and Inspections: Obtain vital allows and agenda inspections to make certain compliance with regulatory requirements.

2. Safety Considerations:

- Emergency Preparedness: Plan for emergency shutoff valves and get entry to factors in case of leaks, floods, or different plumbing emergencies.

- Material Handling: Handle and store plumbing materials, especially chemical compounds and adhesives, consistent with manufacturer instructions and safety tips.

SUSTAINABILITY AND EFFICIENCY

1. Water Conservation:

- Efficient Fixtures: Install water efficient furniture and appliances to limit water consumption and decrease environmental impact.

- Reuse Systems: Consider gray water recycling systems for non potable water uses along with irrigation or lavatory flushing.

2. Energy Efficiency:

- Insulation: Insulate warm water pipes to lessen warmness loss and improve power efficiency.

CHAPTER 4: BASIC PLUMBING REPAIRS

FIXING LEAKS AND DRIPS
CLEARING CLOGGED DRAINS AND BATHROOMS

Clearing clogged drains and lavatories is a not unusual plumbing trouble that can be resolved with the right gear and methods.

Tools and Equipment Needed:

1. Plunger:

- Type: Use a cup plunger for sinks and tubs, and a flange or ball shaped plunger for lavatories.

- Method: Create a seal around the drain opening and plunge vigorously up and down to dislodge the clog.

2. Plumbing Snake (Auger):

- Type: Choose a handheld auger for sinks and tubs, or a lavatory auger specifically designed for lavatories.

- Method: Insert the auger into the drain and crank or push to break up and retrieve the clog.

3. Chemical Drain Cleaners:

- Caution: Use chemical cleaners sparingly and comply with manufacturer commands carefully to avoid damage to pipes and furnishings.

- Method: Pour the advocated amount of purifier into the drain and allow it take a seat as directed before flushing with hot water.

STEPS TO CLEAR CLOGGED DRAINS:

1. Sink and Tub Drains:

- Remove Debris: Clear visible particles from the drain beginning, which includes hair and cleaning soap scum.

- Plunge: Use a plunger to create suction and pressure the clog loose.

- Auger: If plunging doesn't work, use a plumbing snake to reach deeper into the drain and break up the clog.

2. Toilet Clogs:

- Plunge: Use a rest room plunger to create a strong seal around the rest room drain and plunge forcefully to dislodge the clog.

- Toilet Auger: If plunging doesn't work, use a lavatory auger to attain deeper into the toilet lure and break up the clog.

PREVENTIVE MAINTENANCE TIPS:

- Regular Cleaning: Prevent buildup via periodically cleaning drains with a mixture of baking soda and vinegar followed through hot water.

- Hair Catchers: Use hair catchers or displays over drain openings to lure hair and debris earlier than they enter the pipes.

CHAPTER 5: INSTALLATION TECHNIQUES

INSTALLING FAUCETS, SINKS, AND TOILETS

CONNECTING APPLIANCES LIKE DISHWASHERS AND WASHING MACHINES

Connecting home equipment like dishwashers and washing machines to plumbing structures involves specific steps to make certain right functionality and prevent leaks.

DISHWASHER CONNECTION:

1. Location:

- Identify a suitable area close to the kitchen sink where the dishwasher could be hooked up. Ensure there is get entry to to plumbing traces (hot water deliver and drain).

2. Supplies Needed:

- Dishwasher installation kit (normally includes a deliver line, fittings, and drain hose).

- Adjustable wrench or channel lock pliers.

- Screwdrivers (if wished for hose clamps).

3. Steps to Connect:

Water Supply:

Shut off the principle water deliver to the kitchen.

- Connect the dishwasher's warm water deliver line to the hot water shutoff valve underneath the sink the use of the protected adapter (often a compression fitting).

- Tighten connections with an adjustable wrench, making sure they're steady however not over tightened to keep away from damage.

Drain Connection:

- Locate the dishwasher drain outlet at the unit.

- Connect the dishwasher drain hose to the sink drain tailpiece or garbage disposal inlet.

- Secure the hose with a hose clamp to prevent leaks. Ensure the drain hose is looped up to save you back flow of wastewater.

Electrical Connection:

- If putting in a brand new dishwasher, connect the electrical wiring in line with the producer's commands. This typically includes connecting the dishwasher's strength wire to an outlet underneath the sink or hard wiring it to an electrical circuit.

4. Testing:

- Turn at the water deliver and take a look at for leaks at all connections.

- Run a check cycle at the dishwasher to ensure it fills with water, drains properly, and operates without leaks.

WASHING MACHINE CONNECTION:

1. Location:

- Choose a vicinity near a water deliver line (normally a chilly water line) and a drain for the bathing system.

2. Supplies Needed:

- Washing gadget hoses (typically covered with the system or to be had separately).

- Adjustable wrench or channel lock pliers.

- Screwdrivers (for hose clamps).

3. Steps to Connect:

Water Supply:

- Shut off the main water supply to the bathing system place.

- Connect the showering system's hot and cold water hoses to the corresponding hot and bloodless water shutoff valves the use of the blanketed washers.

- Tighten connections with an adjustable wrench, ensuring they're secure.

Drain Connection:

Locate the bathing system drain hose.

- Connect the drain hose to a standpipe or laundry sink drain the use of a hose clamp.

- Ensure the drain hose is secured and placed efficiently to prevent leaks and avoid kinks.

Electrical Connection:

- Plug the washing gadget electricity twine right into a grounded electrical outlet.

- Follow manufacturer instructions for any extra electrical setup or configurations.

4. Testing:

- Turn at the water deliver and test for leaks at all connections.

- Run a take a look at cycle on the washing gadget to make certain it fills with water, drains nicely, and operates with out leaks.

SAFETY TIPS:

- Check Connections: Periodically inspect hose connections for symptoms of damage or leaks.

- Use Proper Drainage: Ensure washing gadget and dishwasher drain hoses are secured and located effectively to save you water damage.

- Follow Manufacturer Instructions: Always refer to the equipment manufacturer's installation suggestions for precise instructions and safety precautions.

CHAPTER 6: MAINTENANCE TIPS

PREVENTATIVE MAINTENANCE STRATEGIES

SEASONAL PLUMBING EXAMS

Seasonal plumbing checks are vital to hold the functionality and performance of your plumbing system in the course of the year.

SPRING AND SUMMER PLUMBING CHECKS:

1. Inspect Outdoor Faucets and Sprinkler Systems:

- Check for leaks, cracks, or damage in out of doors faucets (hose bibs) and irrigation structures.
- Ensure sprinkler heads are functioning nicely and regulate for correct coverage.

2. Check for Leaks:

- Inspect visible pipes below sinks, round bathrooms, and in basements or crawl areas for signs of leaks (e.G., water stains, dampness).

- Repair any leaks right away to prevent water damage and conserve water.

3. Test Sump Pump:

- Pour water into the sump pit to make certain the pump activates and discharges water well.

- Clean the sump pit and put off any particles which can intrude with pump operation.

4. Inspect Water Heater:

- Flush the water heater to eliminate sediment buildup which can lessen efficiency and shorten its lifespan.

- Check the temperature placing and modify if essential (commonly around a hundred and twenty tiers Fahrenheit for protection).

5. Check Toilet and Faucet Functionality:

- Test bathrooms for proper flushing and test for leaks across the base or tank.

- Inspect faucets for drips or leaks and replace worn washers or seals as needed.

FALL AND WINTER PLUMBING CHECKS:

1. Winterize Outdoor Plumbing:

- Drain and disconnect lawn hoses, and shut off outdoor water supply valves.

- Insulate outdoor taps and uncovered pipes to prevent freezing and ability bursts.

2. Inspect Insulation:

- Check insulation round pipes in unheated areas together with basements, move slowly spaces, and attics.

- Add insulation as had to save you pipes from freezing in the course of bloodless climate.

3. Test Water Flow and Pressure:

- Run water in all sinks, showers, and tubs to test for adequate water drift and pressure.

- Monitor for any modifications which could indicate pipe obstructions or plumbing problems.

4. Check Drainage:

- Clear debris from gutters and downspouts to make certain right drainage faraway from the muse.

- Clean out floor drains and make sure they're free of obstructions to save you backups.

5. Inspect Septic System (if relevant):

- Schedule a expert inspection and pumping of the septic tank if advocated via the carrier provider.

- Avoid flushing non biodegradable items or chemical compounds which can disrupt the septic device's balance.

ADDITIONAL TIPS:

- Know Your Plumbing System: Understand the format of your plumbing system and locate close off valves for quick reaction in case of emergencies.

- Schedule Professional Maintenance: Consider scheduling annual plumbing inspections and upkeep with a certified plumber to hit upon and cope with potential troubles early.

CHAPTER 7: GREEN PLUMBING PRACTICES

WATER CONSERVATION TECHNIQUES

ECO-FRIENDLY PLUMBING OPTIONS

Choosing green plumbing options can help conserve water, lessen energy intake, and minimize environmental effect.

WATER SAVING FIXTURES:

1. Low Flow Toilets:

- Advantages: Use extensively much less water according to flush in comparison to older models (e.G., twin flush toilets provide distinct flush alternatives for liquid and strong waste).

- Savings: Can store up to 2060% of water used for flushing.

2. Low Flow Shower heads:

- Advantages: Reduce water intake all through showers with out compromising water stress or comfort.

- Savings: Can save up to 50% of water used for showering compared to standard shower heads.

3. Aerators for Faucets:

- Advantages: Mix air with water to maintain robust water pressure whilst decreasing universal water utilization.

- Savings: Can lessen water flow via as much as 50% with out affecting overall performance.

EFFICIENT WATER HEATING:

1. Tank less Water Heaters (On Demand Water Heaters):

- Advantages: Heat water most effective whilst wished, putting off standby electricity losses related to traditional tank water heaters.

- Savings: Can be up to 30% greater power efficient than traditional tank water heaters.

2. Heat Pump Water Heaters:

- Advantages: Use ambient air to heat water, eating less electricity in comparison to electric powered resistance or gasoline water warmers.

- Savings: Can save up to 60% in energy prices as compared to standard electric powered water heaters.

Grey water Systems:

1. Grey water Recycling Systems:

- Advantages: Recycle water from sinks, showers, and washing machines for irrigation or toilet flushing, reducing clean water usage.

- Savings: Can considerably lessen water intake for landscaping.

SUSTAINABLE MATERIALS:

1. Recycled and Eco Friendly Plumbing Materials:

- Advantages: Use substances crafted from recycled content or sustainable sources, along with eco-friendly piping substances like PEX or HDPE.
- Savings: Contribute to lowering environmental impact via preserving assets and decreasing carbon footprint.

RAINWATER HARVESTING:

1. Rainwater Collection Systems:

Advantages: Collect rainwater from roofs or surfaces for non potable makes use of which include irrigation, landscaping, or lavatory flushing.

Savings: Reduce reliance on municipal water components and preserve freshwater resources.

LEAK DETECTION AND PREVENTION:

1. Smart Leak Detection Devices:

- Advantages: Monitor water usage and hit upon leaks in real time, making an allowance for set off maintenance and water conservation.
- Savings: Prevent water waste and reduce potential water harm to assets.

Eco Friendly Plumbing Practices:

- Regular Maintenance: Schedule habitual inspections and protection to ensure plumbing structures are running successfully and to detect leaks or inefficiencies early.
- Educational Programs: Participate in water conservation applications or workshops to learn about sustainable water use practices and technology.
- Government Incentives: Check for rebates or incentives provided by way of neighborhood governments or software organizations for putting in eco-friendly plumbing fixtures and systems.

CHAPTER 8: LEGAL AND CODE CONSIDERATIONS

UNDERSTANDING PLUMBING CODES AND RULES

PERMITS AND INSPECTIONS

Permits and inspections are critical elements of the plumbing method, ensuring that installations and modifications follow building codes, protection standards, and environmental rules.

Permits:

1. When are Permits Required?

● New Installations: Any new plumbing installations, which includes including new furnishings or appliances (e.G., sinks, bathrooms, water heaters).

● Modifications: Significant modifications or changes to existing plumbing structures, consisting of rerouting pipes or converting pipe sizes.

- Repairs: In some jurisdictions, allows can be required for essential repairs, specifically those involving structural adjustments or sizable pipe replacements.

2. Who Obtains Permits?

- Homeowners: In a few instances, homeowners might also observe for allows themselves, particularly for minor projects. However, for extra complicated installations or modifications, it's often really helpful to lease an authorized plumber who can handle permit programs in your behalf.

3. Permit Application Process:

- Local Regulations: Check with your neighborhood building branch or municipality to determine precise permit requirements, software methods, and fees.

- Documentation: Prepare vital files which includes plumbing plans, specifications, and product statistics as required by means of the constructing branch.

INSPECTIONS:

1. Purpose of Inspections:

- Compliance: Verify that plumbing work meets constructing codes, protection requirements, and nearby policies.

- Safety: Ensure installations are secure and well hooked up to prevent capacity dangers which includes leaks or structural damage.

2. Types of Inspections:

- Roughing Inspection: Conducted before concealment of plumbing paintings (e.G., pipes in the back of walls or underneath flooring) to make sure proper installation and alignment according to plans.

- Final Inspection: Conducted after finishing touch of plumbing paintings to verify that every one furnishings, connections, and installations meet code necessities and are functioning efficaciously.

3. Scheduling Inspections:

- Coordination: Schedule inspections with the local building branch or inspector's office in advance. Inspections may additionally need to be scheduled at one of a kind stages of the assignment (e.G., difficult in and final inspections).

- Notification: Ensure that the plumber or contractor notifies you while inspections are scheduled and coordinate access to the assets as wanted.

4. Passing Inspections:

- Approval: Upon passing inspections, the constructing branch will trouble a Certificate of Compliance or comparable documentation indicating that the plumbing work complies with applicable codes and policies.

- Completion: Keep records of all inspection approvals and documentation for future reference, such as for the duration of belongings sales or renovations.

IMPORTANCE OF COMPLIANCE:

- Legal Compliance: Noncompliance with permit and inspection necessities can result in fines, penalties, or delays in venture final touch.

CHAPTER 9: CAREER AND BUSINESS ASPECTS

TRAINING AND CERTIFICATION OPTIONS

STARTING A PLUMBING ENTERPRISE

Starting a plumbing commercial enterprise may be profitable and worthwhile, however it requires cautious making plans, guidance, and adherence to regulatory necessities.

1. Research and Planning:

- Market Research: Assess neighborhood demand for plumbing offerings, competition, and capacity consumer demographics.

- Business Plan: Create an in depth marketing strategy outlining your services, target market, pricing approach, advertising plan, and economic projections.

- Legal Structure: Decide at the legal shape of your commercial enterprise (e.G., sole proprietorship, partnership, LLC) and sign up your business with the suitable authorities.

2. Obtain Necessary Licenses and Certifications:

- Plumbing License: Obtain the specified plumbing license or certification out of your kingdom or local licensing board. Requirements range by means of region however usually include passing an examination and meeting revel in criteria.

- Business Permits: Obtain any vital enterprise lets in and licenses required to function legally in your location, such as a fashionable business license and exchange particular permits.

3. Set Up Your Business Operations:

- Location: Choose a appropriate place on your enterprise operations, whether it's a

domestic office or a business area.
Consider accessibility, storage desires
for gadget, and zoning rules.

- Insurance: Secure business coverage,
  along with legal responsibility insurance
  and employee's repayment insurance to
  shield your enterprise, employees, and
  customers.

4. Equip Your Business:

- Tools and Equipment: Invest in
  wonderful plumbing equipment,
  equipment, and vehicles (if needed) to
  carry out various plumbing obligations
  effectively.

- Inventory: Stock up on plumbing
  materials, components, and substances
  necessary for commonplace maintenance
  and installations.

5. Marketing and Branding:

- Website and Online Presence: Create a
  professional internet site showcasing

your offerings, touch information, patron testimonials, and blog content material associated with plumbing guidelines or advice.

- Local Marketing: Implement local advertising techniques such as on line marketing, search engine optimization (search engine optimization), social media advertising, and networking with neighborhood organizations and owners.

- Branding: Develop a memorable emblem identity, along with a brand, commercial enterprise playing cards, and uniform layout, to construct credibility and recognition on your network.

6. Establish Customer Relationships:

- Customer Service: Provide notable customer support with the aid of being responsive, transparent, and professional in all interactions.

- Referral Program: Encourage glad customers to refer your services to others thru a referral program or incentives.

7. Financial Management:

- Accounting System: Set up an accounting machine to music earnings, costs, and taxes. Consider the use of accounting software program to streamline monetary control duties.

- Budgeting and Pricing: Develop a pricing shape that covers your expenses and generates income at the same time as last competitive inside the market.

8. Compliance and Safety:

- Compliance: Stay updated on plumbing codes, regulations, and protection requirements to make certain compliance in all your plumbing paintings.

- Safety Practices: Implement safety protocols for yourself and your employees to save you injuries and make certain place of work safety.

9. Build a Team (if applicable):

- Hiring: As your business grows, take into account hiring qualified plumbers or apprentices to expand your potential and serve greater clients.

- Training: Provide ongoing schooling and development possibilities to make sure your group remains professional and knowledgeable approximately industry tendencies and nice practices.

10. Grow and Adapt:

- Customer Feedback: Collect and examine purchaser remarks to enhance your services and client satisfaction.

- Diversification: Consider imparting extra services inclusive of HVAC restore, water remedy structures, or emergency plumbing offerings to make bigger your commercial enterprise services.

www.ingramcontent.com/pod-product-compliance
Lightning Source LLC
Chambersburg PA
CBHW072001210526
45479CB00003B/1030